PARTICIPATION IN CHANGE
New Technology and the Role of Employee Involvement

PARTICIPATION IN CHANGE
New Technology and the Role of Employee Involvement

Research Results on Participation in Technological Change

PETER CRESSEY, UNIVERSITY OF BATH
ROBIN WILLIAMS, UNIVERSITY OF EDINBURGH

European Foundation
for the Improvement of
Living and Working Conditions
Loughlinstown House, Shankill, Co. Dublin, Ireland
Telephone: 826888   Telex: 30726 EURF EI   Fax: 826456

This booklet is also available in the following languages:

ES	ISBN 92-826-0233-8
DA	ISBN 92-826-0234-6
DE	ISBN 92-826-0235-4
GR	ISBN 92-826-0236-2
FR	ISBN 92-826-0238-9
IT	ISBN 92-826-0239-7
NL	ISBN 92-826-0240-0
PT	ISBN 92-826-0241-9

Luxembourg: OFFICE FOR OFFICIAL PUBLICATIONS OF THE EUROPEAN COMMUNITIES 1990

Cataloguing data can be found at the end of this publication

ISBN: 92-826-0237-0

Catalogue number: SY-58-90-384-EN-C

© Copyright: THE EUROPEAN FOUNDATION FOR THE IMPROVEMENT OF LIVING AND WORKING CONDITIONS, 1990. For rights of translation or reproduction, application should be made to the Director, European Foundation for the Improvement of Living and Working Conditions, Loughlinstown House, Shankill, Co. Dublin, Ireland.

Typesetting and Print Production:
Printset & Design Ltd., Dublin

Original language: English

Printed in Ireland

Preface

"Consultation of workers", "Involvement", "Participation" — these are all aspects of an issue high on the agenda in the development of the social dimension of the internal market in Europe.

The social partners in Europe, the governments and the European Commission all have shown a genuine interest in developing and increasing the level of involvement in the work places of Europe. Expression to this has been given in a joint agreement in Val Duchesse. The strategies of the involved partners cover a wide range of means and in recent years many different approaches have been tried at European, Member State and Sector level.

The differences in attitudes towards means of increasing the level of involvement are based on complex patterns of national and cultural traditions in industrial relations in the Member States.

To assist the participants in the debate, the Foundation has in its recent programmes of work gathered a range of information pertinent to this issue. Through case studies in public and private sector it has studied factors facilitating involvement in technological change. In an attitudinal survey, it has explored employer and employee representatives' perceptions of the level, and consequences, of different forms and degrees of participation.

In 1988-1990 the Foundation has held a series of national round tables where the social partners have been able to discuss the options and the Foundation's findings.

The present booklet summarises the findings of the Foundation's work as they have been presented to the social partners. It is our hope that this booklet will further stimulate the debate on means of increasing the level of involvement at work places in Europe.

Clive Purkiss
Director

Eric Verborgh
Deputy Director

Contents

Chapter 1	**INTRODUCTION**	
	The Research Background	12
	Why Participation?	13
	Technological change at work	18
	Participation and technological change	21
Chapter 2	**THE FOUNDATION'S RESEARCH PROGRAMME**	
	Methodology	28
	The Research Questions	32
Chapter 3	**THE RESULTS OF THE SURVEY**	
	The Profile of Current Involvement	38
	Why this Pattern of Participation?	43
	The Benefits of Participation	50
	Involvement in the Future	54
Chapter 4	**INTERNATIONAL COMPARISON**	
	Different National Models	64
	Implications of International Comparison	72
Chapter 5	**CONCLUSIONS**	
	Conclusions	76
	Questions for the Social Partners	77
Appendix 1	International Comparative Data on Remaining Six Countries	82
Appendix 2	Relevant European Research Studies	85

List of Tables

Table 1	Management responses to current involvement in planning and implementation	39
Table 2	The paradox of participation	40
Table 3	Management responses to the coverage of participation	41
Table 4	Management responses to the coverage of more intense forms of participation	42
Table 5	Assessment of participation by employee representatives	51
Table 6	Assessment of participation by management	52
Table 7	Involvement currently, and sought in future by management	54
Table 8	Involvement currently, and sought in future by employee representatives	55
Table 9	More intense forms of involvement currently, and sought in future by management	56
Table 10	More intense forms of involvement currently, and sought in future by employee representatives	57

Table 11	Coverage of future involvement by management	58
Table 12	Coverage of future involvement by employee representatives	60
Table 13	International comparison (all forms of participation) current and future participation by management	66
Table 14	International comparison (all forms of participation) current and future participation by employee representatives	66
Table 15	International comparison (more intense forms) current and future by management	67
Table 16	International comparison (more intense forms) current and future participation by employee representatives	67
Table 17	Remaining six countries — Planning and implementation current and future (all forms) by management	83
Table 18	Remaining six countries — Planning and implementation current and future (all forms) by employee representatives	83
Table 19	Remaining six countries — Planning and implementation current and future (more intense forms) by management	84
Table 20	Remaining six countries — Planning and implementation current and future (more intense forms) by employee representatives	84

CHAPTER 1

INTRODUCTION

The Research Background

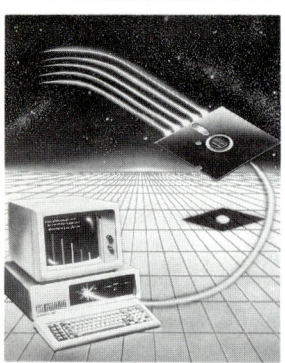

Information Technology, computers and equipment based on micro-electronics have become diffused across all sectors of industry. There can be few workplaces which have not been affected by the introduction of such 'new technology'. There is increasing recognition of the importance of the social and organisational aspects of technological change, as well as their potential significance for the success of change, and for its employment implications. As a result the social partners — trade unions, employers and government bodies — have a continuing concern with policies for new technology in the workplace. One area which has attracted considerable attention is the role of joint participation among employers, the workforce and their representatives in the introduction of new technology.

The European Foundation for the Improvement of Living and Working Conditions set out to inform this debate by conducting research, and through discussions with representatives of the social parties and with expert groups. This report presents the findings of a long-term programme of research initiated by the Foundation into 'Participation and New Technology'. The first set of findings reported here arises from a recent survey, of the attitudes and experiences of managers and employee representatives, covering all the twelve countries within the European Community, and the Foundation's earlier case-study and analytical work.

The Foundation's attitudinal survey was the most comprehensive international survey of these issues to date. It sought to establish and explain the current practice and patterns of participation over technologies currently being introduced in European enterprises. In addition, the survey sought to address the concerns of the social partners in developing future policy. The survey thus asked the managers and employee representatives directly involved in technological change to assess the influence of participation, and to evaluate its benefits and costs. It further asked respondents whether, and in what manner, they would seek to engage in participation over technological change in future.

This report draws upon survey evidence, case studies and analysis of other Foundation research to inform debate on participation and new technology. Can participation contribute to the effective and harmonious implementation of new technology? How can participatory practices be enhanced and improved? The research programme also focused on the wide range of participatory practices that exists across Europe; these vary according to the particular national context as well as the specific characteristics of the industrial sector and enterprises concerned. This report examines participation at a European level, as well as particular national experiences. In this way it addresses the policy concerns of the social partners at both national and European level.

A large volume of data has been collected and is summarised in this report. But before we go on to present and discuss these findings, it is useful to examine the background.

Why Participation?

During the extensive research programme initiated by the Foundation, some effort has been made to understand why the different parties in industry find participation important, why they are interested in the topic and what reasons and rationales for participation are most often given. The Foundation's case studies provided a number of

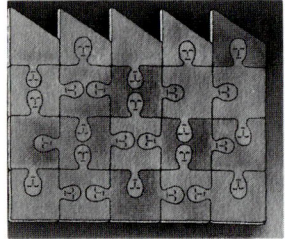

different reasons for participation.

☐ *Efficiency:* participation was seen as an aid to corporate effectiveness, to enable cost savings through the rational and agreed use of resources.

☐ *Motivation* was mentioned frequently; here participation unlocks the problem-solving ability, human resources and creativity of the workforce

☐ *Redistribution of Power:* Some actors identified a power imbalance inside enterprises. Participation was a means of redressing that balance and leading to a more equitable distribution of resources and control of the process of change.

☐ *Good industrial relations:* Participation was seen to be concerned with maintaining amicable relations in the enterprise, possibly as part of a management strategy aimed at encouraging an employee's willingness to take on responsibilities by becoming identified with the company objectives. Participation may, in this sense, be a vehicle for good communications in the enterprise.

We see from this the many ways in which the term participation was used and how it reflected a range of different objectives and rationales for participation. The actors, however, did not necessarily choose only one of these; they indicated both primary and secondary concerns.

The facing illustration demonstrates that the social parties gave priority to different sets of issues. Managers valued the role of participation where it directly contributed to efficiency, flexibility and

CONCERNS REGARDING PARTICIPATION

Primary Concerns of MANAGEMENT	Efficiency Motivation Facilitation of Change
Primary Concerns of WORKFORCE	Power Balance Control of Impacts
Secondary Concerns of BOTH PARTIES	Commitment Communications Good Industrial Relations Identification

motivation. Participation here aided and facilitated change and offered substantial benefits in the form of improved productivity and work procedures.

The workforce and trade union representatives emphasised the democratic role of participation in redressing power imbalances within the enterprise. They were concerned with opportunities, afforded by various forms of participation, to influence the decision-making processes surrounding organisational change. Furthermore, participation could also provide agreed protection against possible negative impacts of change.

It might be thought that the case studies undertaken by the Foundation show only two distinct models of participation emerging; in reality, the situation was more complex. In addition to their primary concerns, the parties expressed a series of secondary concerns; these in practice were often shared by both parties. For instance, identification with, and a commitment to, the success of the enterprise was a goal shared by many. They agreed on the need for appropriate industrial relations procedures, for a good and safe working environment and for a better system of communications. In these areas, agreement was possible and participation did take off, as the processes of problem solving and accommodation it afforded satisfied the interests of both parties.

We can detect three broad models of participation. In addition to the traditional dualistic perspective on participation, as either a productive force, "a management tool to gain increased efficiency" or as a democratic force "for redistributing influence", we see a third model, of participation as a positive-sum game. The last has received increased attention in recent years.

Models of Participation
These three models of participation are discussed below.

Participation as a Productive Force
In this sense participation is seen as directly contributing to the successful implementation of change, and to securing the commercial objectives of the organisation. The managers and the workforce appreciate the problems caused by the artificial separation of production issues on the one side, and the human resource issues on the other. In a situation of change and uncertainty, these latter factors can prove decisive and determine the degree of success that the enterprise actually enjoys. Participative measures in these circumstances aim at securing the consent and commitment of the workforce to new working arrangements, and can include initiatives which directly engage employees in planning and implementing change. Such initiatives can be seen in many of the case studies in

the form of joint working parties, direct worker input into problem-solving and other structured fora that aid management in the act of managing.

Participation as a Democratic Force
This approach to participation is primarily orientated towards decision-making processes in the organisation. The prime objectives include the equalisation of representation of the different parties, the minimisation of conflict and the achievement of formal agreements on the actors' approach to change. Such participation often forms part of the formalised system of representation in the organisation, and may be sustained by legislation, by sectoral or national contracts, or by agreements between the social partners. This approach to participation is directly influenced by the particular industrial relations system involved; it is assumed that an industrial relations situation will be improved only if employees and their representatives have access to all levels of decision-making, and can successfully protect workers' rights during technological or organisational change. Fundamentally it assumes a zero-sum picture of power, inasmuch as any increase in the worker representative's influence will be at the cost of a diminution of management's influence.

Participation as a Positive-Sum Game
Participation as a positive-sum game typically reflects elements of both the above approaches. The primary concerns of the parties in participation are clearly different, so difficulties can be expected to arise in participation. However, participation may be successful for both parties, as the actors share an array of secondary concerns. This model of participation does not necessarily challenge the

above models; instead it shows how both sides of industry trade-off their interests, have scope to compromise, seek to avoid head-on conflict over points of principle, and pursue objectives that may yield 'win-win' results and minimise the negative consequences for either side. Participation could fulfil the mutual interests of both parties — it could thus be a positive-sum game between the parties.

Technological Change at Work

The Foundation's research, through both the case study and the survey stages, has looked at participation in relation to technological change. From the outset, this raises two important questions:

☐ does the process of technological change allow scope for, or detract from, the development of participation?

☐ what is the contribution of participation to the successful and balanced introduction of new technologies?

The initial debate on the implications of new technology at work painted a problematic picture. Many of the early commentators predicted that the introduction of micro-electronics and information technology (IT) would be rapid and widespread, and would have dramatic "impacts" on employment levels and on skill requirements. These writings were often based upon "technological-determinist" assumptions that if a particular technology were to be introduced, then predetermined results would inevitably follow in the areas of skill, job numbers, working conditions and operator control. Simplistic forecasts were made about changes in employment levels and working conditions. These stimulated widespread public concern about potentially negative aspects of technological change.

tasks and existing separate systems within the organisation become integrated; and as their effects across the organisation become more diffuse. The need for such involvement is in many instances a direct response to the uncertainty that accompanies innovation.

UNCERTAINTIES IN TECHNOLOGICAL INNOVATION

PROCESS	OUTCOME
How to manage the introduction of new technologies: – **PLANNING** – **IMPLEMENTATION** – **OPERATION**	**What will be the** – **ECONOMIC** – **ORGANISATIONAL** – **EMPLOYMENT** – **SOCIAL** **consequences of technological change?**

Information Technology is a highly flexible "enabling technology". This feature underpins its potentially widespread application, the scope for choice in the way it is used and the outcomes for the organisation. As a result, its implementation is often accompanied by a high level of uncertainty amongst the actors. The growing pace of technological change augments this uncertainty, because of the increasing complexity and novelty of many applications. Technological change results in a double uncertainty:

Participation and Technological Change

There is considerable scope for joint discussion and participation over the introduction of new technologies, and it includes a host of key issues such as productivity, hardware design, work organisation and training. The range of issues demonstrates that participation is not restricted to minor or peripheral aspects of the innovation process, but can be central.

SCOPE FOR JOINT DISCUSSION AND PARTICIPATION

- **Work Organisation and Flow**
- **Hardware and its Configuration**
- **Design of Software and Systems**
- **Productivity**
- **Quality of Products**
- **Health and Safety**
- **Employment Security**
- **Training and Skill**
- **Job Evaluation and Remuneration**

With increasing experience of technological change, new approaches to its implementation are emerging. These take into account the wide range of choices facing the organisation during periods of technological change, and their potentially important impacts on groups within the organisation. Increasingly, technological change is no longer seen as the exclusive province of the specialist or technician. Instead, the issues that surround the development of technological systems are seen to require the involvement of an extensive range of actors. Such involvement becomes ever more necessary as IT-based systems become more comprehensive; as they are applied to more complex

There are no simple (eg economic or technical) criteria for the successful design of new technology systems. Establishing which particular solution would be most successful and appropriate for a given organisation depends on many factors, including the technical features of the production process, batch size, product market and dynamics, plant size, the structure and qualification of the workforce, local labour markets, traditions, and institutional structures at plant, industry and national levels. Managerial, professional and occupational groups often differ about the most effective way to proceed.

The existence of this complex array of choices opens up an arena for discussion by the different actors involved. The range of concerns to be addressed includes: the technical and organisational options; their implications for different groups and for the various goals of the organisation; finding the best way of balancing these different considerations. Successful management of technology can depend on the ability of organisations to weld these potentially divergent perspectives into a coherent effective and acceptable whole, and participation between the social parties can make an important contribution here. This opens up the potential for social dialogue at the level of the workplace.

During this initial debate on the employment effects of new technology, there was a marked polarisation between pessimists and optimists as they discussed whether job displacement would be balanced by new job creation, whether destruction of old skills would be balanced by the creation of new skills, and whether forms of job control and working conditions would improve or worsen.

In its research the Foundation chose not to enter that debate, and instead focussed on the opportunities for influencing the process of technological change, and its economic and social outcomes. The Foundation's research examined how open the process was and what space it gave for employee participation.

Cumulative research findings soon demonstrated that technological change does not have inevitable and predetermined "impacts" on organisations. In the introduction of any given technological innovation, there exists scope for a variety of solutions in terms of work organisation and in the design and configuration of technology.

The Scope for Choice in Technological Change

Various choices are available at each stage of technological innovation, in terms of both the technical and organisational solutions adopted. Choices arise in the selection of hardware, in the design and configuration of software and systems and in the organisational arrangements for the use of technological systems. These choices have important implications for the rate and success of technological change and its outcomes for the social parties — the commercial benefits, employment levels, the division of labour and skills, and the quality of working life.

> ' Technology does not pre-determine impacts. Instead, technology allows for a variety of solutions, and a range of technological and organisational choices. '

☐ the process of introducing change, ie how to manage the planning, implementation and operation of the technology, involves difficult questions for managers about the most appropriate procedures for managing change.

☐ the outcomes of the change, the economic employment, organisational and social consequences, have to be substantively resolved to the satisfaction of all parties. Such issues proved to be the source of much conflict and anxiety in many of the case studies.

A Role for Participation in Technological Change

Participation could minimise the uncertainties in both of these areas, aiding the process of design and the implementation of changes. This could be through joint activity in problem-solving, or by helping to articulate organisational needs and objectives for new technological systems. In the area of outcomes, participation was seen as a means of

A ROLE FOR PARTICIPATION

PROCESS	OUTCOME
Participation as an aid to joint problem solving in the design and implementation of technological change ↓ allows for workforce influence	Participation as a means of balancing impacts and alleviation fears ↓ facilitates organisational adaptation

balancing the impacts and at the same time ending rumours and thereby alleviating any unfounded fears regarding the impact of the change.

These rationales for participation were not in practice distinct, but could be mutually reinforcing. Thus, participation which minimised negative outcomes (and fear of them) facilitated adaptation by organisation members. Equally, direct involvement in problem-solving opened opportunities for workforce influence over the outcomes of change. Participation offered a two-dimensional process, allowing worker involvement in the actual outcomes, and aiding the setting up of adaptive mechanisms and procedures for introducing innovation.

The complexity, indeterminacy and uncertainty that emerged in the case studies, and in the later research, suggest a view of the process of technological change as being one of organisational learning and organisational adaptation, as well as technological problem-solving. This perspective emphasises the potential contribution of a wide range of members of an organisation to the successful implementation of change. Successful outcomes depend on the contribution of skills, expertise, experience and ingenuity by users as well as by technical and other specialists.

Such a new perspective is recognised in such initiatives as:

- user involvement in computer-system development;
- project groups and Quality Circles;
- new technology committees;
- ad hoc project groups;

☐   the range of consultative and joint decision-making fora set up to deal with the co-ordination of human resources during the introduction of change.

However, uncertainties also pose problems for participation. In particular, it is difficult for organisation members to assess technological opportunities and predict their implications for the organisation. This is a problem for all parties to resolve — including managers and engineers — but it is particularly acute for those who lack technical expertise.

In summary, we have looked at the reasons for participation and identified a set of positions, problems, and opportunities for it. This has been at the root of the Foundation's concern with participation and technological change since the beginning of the 1980s. There has also been a resurgence of interest in these issues within Europe, culminating in the recent Val Duchesse Joint Opinion on information and consultation during the introduction of technological change. We now turn to the Foundation's research programme.

CHAPTER 2

THE FOUNDATION'S RESEARCH PROGRAMME

Methodology

One of the Foundation's aims has been to document the varieties of participation and regulation that exist within a range of European enterprises while new technology is being introduced. It has done so by using a variety of methods including:

— case-studies
— expert analysis
— attitudinal surveys.

The first phase of the Foundation's research involved detailed case-studies of the involvement of parties in the introduction of new technologies. This involved 21 case studies across five countries with a sectoral spread that included banking, printing, mechanical engineering, electronics and telecommunications. The case studies afforded a detailed examination of participation processes and of the processes of managing the introduction of technological change. A second stage analysis of participation and technological change drew upon these cases and an additional 43 European cases. This phase of research extended both the national and sectoral spread of the case studies; an account of this phase can be found in the Foundation's publication, 'Participation

PARTICIPATIVE MECHANISMS

NO INVOLVEMENT	Management-planned & executed schemes Unilateral management control
INFORMATION PROVISION	Briefing sessions Information agreements Group fora on change
CONSULTATION	Advisory Committees Project groups/new technology Committees/Steering groups
NEGOTIATION	Productivity Agreements Protective clauses in general agreements Planning agreements/N.T. agreements
JOINT DECISION MAKING	Veto Powers Status Quo clauses Joint Decision Bodies

Review'. This was followed by another phase of expert analysis, and then by the attitudinal survey. This final stage of the research was motivated by a concern to generalise about these phenomena and to establish the scope for increasing participation.

The major feature of this survey was its breadth of coverage. The attitudinal survey was concerned to describe the contours of participation found in European enterprises. It did not seek a random sample; rather it concentrated on those companies where participation during technological change had been used in the past. It covered all 12 Community countries. Interviews were conducted with 7326 respondents, comprising equal numbers of management and employee representatives. Management respondents were personnel and line managers. The initial contact with employee representatives was established through management. The survey focused on cases where new technology was having significant implications for work, and concentrated on information technology, computer-based systems and equipment incorporating micro-electronics. A range of industrial sectors was selected:

— *in manufacturing,* mechanical engineering and electrical engineering were represented,

— *in the service sector,* retail, banking and insurance provided the research focus.

The approach of the Foundation's research was agreed with the social partners represented on its Administrative Board. It concentrated on participative mechanisms which were primarily representative in character, involving either workforce or trade union constituencies.

The Foundation's *Participation Review* identifies

' The meaning and the political significance of the term "participation" differ between countries and industrial relations systems. '

many different forms of participation across Europe. This is partly a consequence of the diversity of industrial relations systems, with their specific national traditions, structures and approaches. Moreover, the meaning and the political significance of the term "participation" differ between countries and industrial relations systems. The starting point for the research was the recognition that a wide variety of practices exists in relation to technological change, all of which could be seen as participatory. In order to avoid designating any one model of participation as correct, it was agreed that the Foundation's survey should document all the ways in which the parties achieve involvement in the introduction of new technology.

The Range of Participative Mechanisms

Involvement was defined as any participatory procedure or practice that, formally or informally, directly or indirectly, involves the parties concerned in the discussion of decisions concerning the process of change. This inclusive approach was adopted in accordance with the wishes of the social partners. Such involvement covered a wide range of practices,

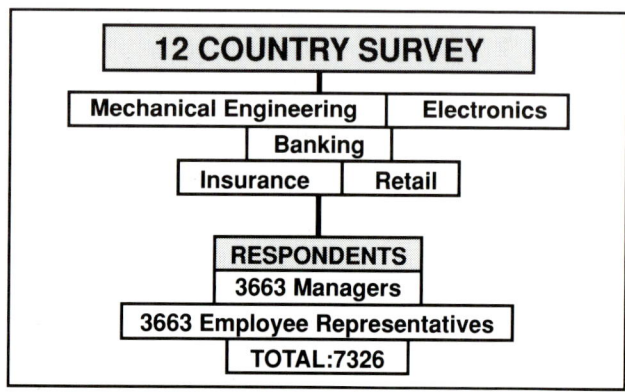

ranging from disclosure of information to consultation, negotiation and joint decision-making. These different forms of participation are grouped according to the degree to which the two parties engaged in joint dialogue and were committed to reaching an agreed solution.

Phases in the Process of Technological Change
Design of the attitudinal survey was influenced by the earlier case-study analysis. Two key concepts concerned the range of practices that go to make up participation and, importantly, the different phases that go to make up the process of introducing new technology. These phases are: planning, selection, implementation and post evaluation.

PHASES IN THE INNOVATION PROCESS

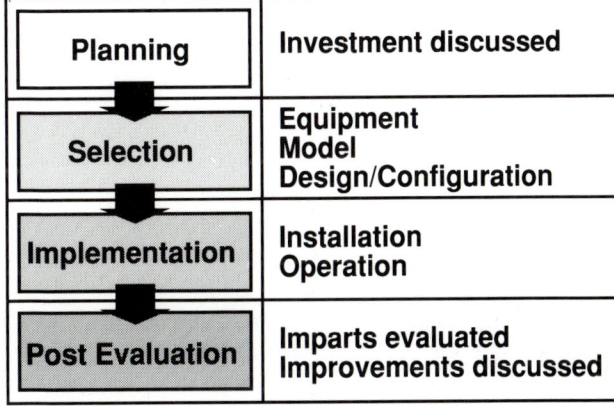

Planning	Investment discussed
Selection	Equipment Model Design/Configuration
Implementation	Installation Operation
Post Evaluation	Imparts evaluated Improvements discussed

These phases are not equal in terms of their importance to the decision-making process. Indeed the strategic decisions regarding change were often made in the early planning and selection phases, with some irreversible effects for the phases further down the chain — the operational phases of implementing and monitoring change. The room for manoeuvre was seen to be reduced as the different phases of introducing change were gone through. After the planning phase the scope for discussion about the broader strategic issues is sharply reduced; choice becomes progressively limited with the investment of time and money in particular technological options.

The Room for Manoeuvre in Technological Change

The cost and difficulty of adapting systems to the requirements of organisation members become greater as one moves through the selection phase and towards implementation. Early discussion of proposed technological change could minimise the costs of adjustment and thereby facilitate the introduction of change in an equitable and harmonious manner. The *timing* of participation is therefore of particular significance.

The Research Questions

The case-studies and the attitudinal survey both sought to establish the character of participatory practices, and their consequences for decision-making and the implementation of technological change. The attitudinal survey in addition asked respondents to evaluate current participatory practices, and to indicate how they would like participation to take place over future technological changes.

The case studies revealed that participatory practices were very uneven — varying in their extent between different phases of the introduction of technological change, and varying in their coverage of different aspects of decision-making and change. Participation tended to be poorly developed during the advance planning of technological change, and around strategic management decisions. Instead, participation tended to focus on dealing with the employment and operational consequences of these decisions, at the stage at which technology was being implemented.

The attitudinal survey provided a systematic account of the overall pattern of participation over technological change. It paid particular attention to three dimensions: the timing, the scope and the intensity of participation.

TIMING — the survey tracked changes in the form and practice of participation through the different phases of technological change.

SCOPE — the survey examined the coverage of different issues in technological change. The survey established the extent to which participatory practices covered a range of issues — from management strategies to operational employment issues.

INTENSITY — the survey examined not only whether employee representatives were involved during different phases and around different issues, but also the manner and degree of their involvement. The survey addressed the level of involvement — the proportion of respondents reporting the operation of participation — and the intensity of this participation.

An extensive set of data has been assembled. In order to facilitate the analysis and presentation of this vast body of findings, it is necessary to condense these data and concentrate on the key variables.

In analysing the timing of participation, this report highlights the differences in the extent of involvement between the advance planning of change, and the implementation of change. Contrasting these two phases captures the most significant features of the timing of participation.

As we show below, the various forms of participation were not employed to the same degree during the stages of the innovation process. There were notable differences in the level and intensity of involvement between the planning stage and the implementation stage. There were also significant differences in the extent of the coverage of strategic, operational and employment aspects of change — differences in the frequency of cases in which participation covered a particular issue, and differences in the manner and intensity of such involvement.

We use here two measures of the extent of participation. First, we look at the number of cases in which respondents reported the use of any of the participatory practices. This is referred to as "all forms" of participation. We then distinguish those cases in which "more intense forms" of participation were adopted. This refers solely to the cases where participation involved negotiation and joint decision-making. The figures for "all forms of participation" also include those cases where participation took the form of information provision and consultation. Contrasting these two measures of participation highlights the basic pattern of participation and its distribution.

Throughout the survey, the replies of managers and employee representatives were remarkably consistent. This is particularly the case when current practices are being described. This enhances our confidence in the validity of the responses. Differences in response are most notable where the social parties were asked about their future expectations for participation and, to a lesser extent, in their evaluation of current practices. We will note areas in which there were significantly different responses by the parties. In order to reduce the volume of data to be displayed, we will not show responses of employee representatives where these merely duplicate management responses.

CHAPTER 3

THE RESULTS OF THE SURVEY

The first objective of the attitudinal survey was to document the overall pattern of participation in technological change. It sought to answer the following questions:

- [ ] how did participation vary between different phases of technological change?
- [ ] how did the coverage of participation vary between different aspects and issues?
- [ ] did participation take place around particular phases?
- [ ] what was the manner and intensity of involvement?

To answer these questions we now turn to the results of the Attitudinal Survey, addressing first the findings at a European-wide level.

The Profile of Current Involvement

The first part of the attitudinal survey examined the participation practices that were currently being adopted. Managers and employee representatives were asked what participation took place, at different phases of change and around different issues.

Table 1 shows the responses of Managers about the extent of involvement of employee representatives. They were asked to categorise the type of participation practice adopted.

The overall pattern shows that the planning phase tended to be characterised by involvement which was limited to the least intense form, that of information disclosure (79%). As one moves towards the implementation phase, there is a marked increase in the overall frequency of involvement. In particular, there is a shift from information

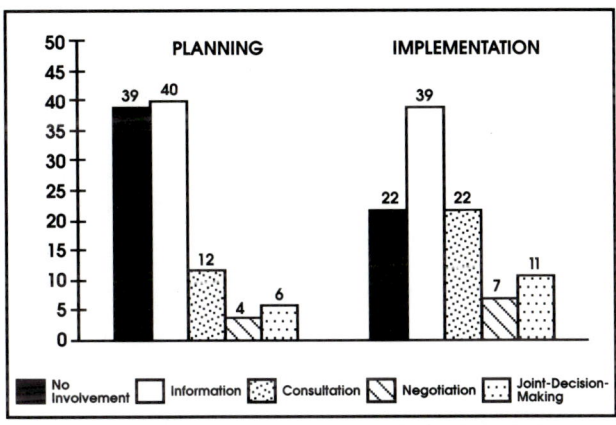

TABLE 1: MANAGEMENT RESPONSES TO CURRENT INVOLVEMENT IN PLANNING AND IMPLEMENTATION

	No Involvement	Information	Consultation	Negotiation	Joint-Decision-Making
Planning	39	40	12	4	6
Implementation	22	39	22	7	11

provision to consultation, and an increase in the frequency of more intense forms of participation (negotiation and joint decision-making) over the technology to be introduced. This finding repeated the pattern found in the case studies. The pattern of involvement — showing a rise in intensity during the stage of implementation and a lower frequency/intensity during the earlier stages — is remarkably consistent across the whole sample. The evidence highlights a number of subtle patterns and variations in the distribution of participation. This pattern was also found to exist across the different national situations.

The Paradox of Participation

When we refer back to the room for manoeuvre during the process of technological change, we find an intriguing picture. As the scope for influence decreased during the innovation process, so the intensity of participation increased. It appears that an increase in participation takes place as the chance to exercise significant influence diminishes.

Table 2 illustrates this paradox of participation.

For the purposes of this study we will concentrate mainly on the comparison between "all forms" and "more intense forms" of participation in order to see clearly how the different intensities of participation manifest themselves across the phases.

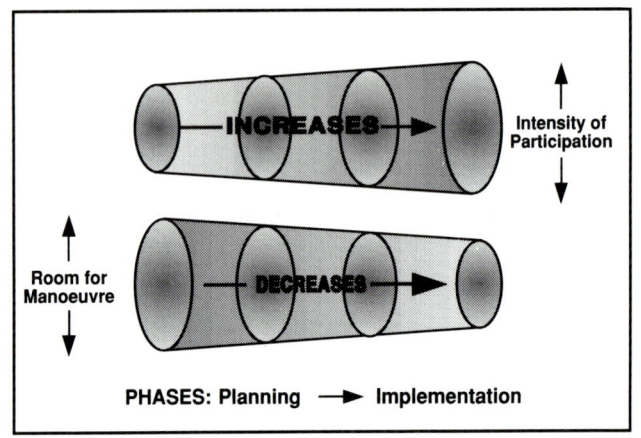

TABLE 2: THE PARADOX OF PARTICIPATION

The pattern found in relation to the timing and intensity of participation is also supplemented by other evidence from the attitudinal survey, which looks at the different issues that are covered by participatory practices. The survey examined the *content* of participation — whether participation took place around a range of issues.

Issues Covered in Participation
Table 3 shows the frequency with which management reported involvement in a range of issues. The issues on the left of the table are primary management concerns — strategic concerns followed by operational concerns. The middle of the

	ALL FORMS OF INVOLVEMENT
Primary workforce concerns	HEALTH AND SAFETY — 81
	JOB SATISFACTION — 76
	TASK SPECIFICATION — 81
	WORK ORGANISATION — 81
	PRODUCT INTRODUCTION — 71
	REDUCING COSTS — 59
	INVESTMENT CRITERIA — 45
Primary management concerns	MARKET STRATEGY — 45

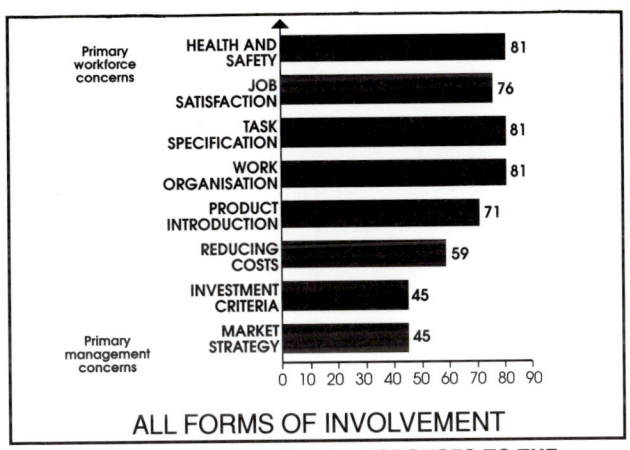

TABLE 3: MANAGEMENT RESPONSES TO THE COVERAGE OF PARTICIPATION

table represents the joint concerns of both management and workforce, whilst the issues to the right of the table are the particular concerns of the workforce. The table shows the coverage of involvement to be weighted to the right hand side of the table. Involvement seems to have been *least* frequent in those areas where management has traditionally exercised a prerogative, particularly in the strategic concerns regarding product and market strategy and investment criteria. In contrast, involvement was markedly *more* frequent on issues of joint concern and issues close to the everyday problems of running an enterprise — work organisation, task specification and health and safety. The same distribution occurs in both the management and workforce responses. This indicates that there was a boundary around the areas where participation took place. This is similar to, and supplements, the pattern of restricted participation in the earlier phases of the planning process, in which strategic issues are decided, and more extensive participation in the implementation phase, where the actual organisation of the innovation is hammered out.

Category	Issue	Value
Primary workforce concerns	HEALTH AND SAFETY	33
	JOB SATISFACTION	24
	TASK SPECIFICATION	27
	WORK ORGANISATION	26
	PRODUCT INTRODUCTION	13
	REDUCING COSTS	12
	INVESTMENT CRITERIA	7
Primary management concerns	MARKET STRATEGY	6

ALL FORMS OF INVOLVEMENT

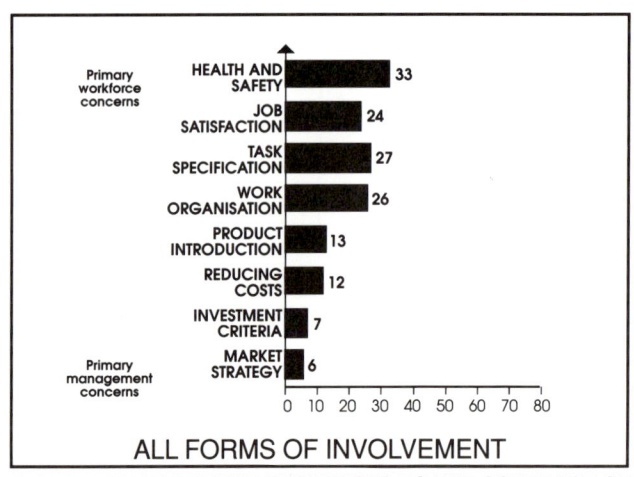

TABLE 4: MANAGEMENT RESPONSES TO THE COVERAGE OF MORE INTENSE FORMS OF PARTICIPATION

When we examine the coverage of different issues by more intense forms of participation, the lines of managerial prerogative become much sharper. As *Table 4* shows, intense forms of involvement made up only a sixth of all forms of involvement in strategic management concerns with markets and investment. By contrast, intense involvement made up about one-third of all forms of involvement over joint managerial/workforce concerns such as task specification and health and safety. Intense involvement was thus even more skewed away from management strategy. In other words, even where managers allowed employee representatives to become involved around traditional areas of managerial prerogative, this involvement was low in intensity and tended to be informal.

These findings reinforce the earlier evidence about the limited character of involvement. It would appear that management was willing to accept some form of employee representative involvement, but sought to minimise the possibility of influence by the workforce. This was evidenced by the relative exclusion of employee participation (especially in its more intense forms) from the strategic early phases of planning technological change. In addition, the participation that did occur covered mainly the operational and employment aspects of change, at the expense of broader aspects of management strategy and policy formation. Even where employee representatives were involved in the early phases, and over strategic issues, this involvement tended to be weak and less formalised.

Why this Pattern of Participation?

The overall evidence raises some key questions about the accepted boundaries of involvement and the apparent restriction of any real opportunities for influence over decision-making at the early phases. Why should such a pattern occur?

Why Participation was Low in the Planning Phase: Managers' Perspectives (M)

M management expressed concerns about the possible influence of participation on the decision-making process for new technologies.

☐ Managers saw the planning of technological change as a management function, and expressed fears over their loss of "managerial prerogative".

- [ ] Some managers suggested that the involvement of employee representatives could potentially give rise to breaches of commercial secrecy.

- [ ] It was felt that elaborate and protracted advance discussion with employee representatives might delay crucial decision-making. Equally the involvement of a broader range of parties would increase the range of objectives for new projects. To the extent that employee representatives suggested additional priorities, management objectives could become submerged or compromised. Managers found the planning of change a difficult process, and one that was beset by uncertainties. They were concerned that employee involvement here would increase uncertainty.

- [ ] Features of the existing decision-making processes were unfavourable to employee representative involvement. This could be a problem both where management decision-making was highly centralised, and also where the processes of planning and design were highly diffuse.

- [ ] Employee representatives were seen as lacking the necessary resources and skills for participation on technology. And there were few legal requirements to involve representatives in the earliest stages of change.

Why Participation was Low in the Planning Phase: Employee Perspectives (E)

E   Employee representative concerns largely revolved around the fact that such involvement creates role confusion for representatives, most acutely in the early stages of decision-making.

☐   Employee representatives expressed particular concern about the hostage phenomenon — whereby their involvement in decision-making might give them responsibility for the outcomes of technological change, without affording them real power to influence technological change and in particular its employment consequences.

☐   Employee representatives were concerned that they were not represented at the management level that was making the key decisions. This in part was seen as a result of management resistance to involvement at the relevant stage, and in part to a lack of fit between representative structures and management decision-making structures.

☐   Employee representatives were concerned that they lacked the resources needed to underwrite involvement, including access to technical information and expertise, and the time needed to examine management proposals and to suggest alternatives. They had little understanding of the process of technological change and the range of technical opportunities available. They tended to downplay their abilities to influence management decisions in this arena. Moreover, employee representatives often did not have an alternative model for how change might be introduced. They needed alternative criteria for assessing new technologies, and practical examples of how these worked. Where employee representatives had failed to articulate their own perspective on change, they lacked the confidence to challenge management plans.

☐ Employee representatives pointed to the fact that there were few or no legal requirements to consult at the early stages.

☐ However, many employees and their representatives were broadly satisfied with the overall process of technological change — either because its impacts upon the workforce were broadly positive, or because it was seen as inevitable. Frequently, employee representatives were convinced by the technologically deterministic arguments promoted by many managers and suppliers that there was only "one best way" of introducing change — that there was an economic or technological imperative to introduce particular types of innovation. This reflects the difficulties for employee representatives in establishing an alternative view of technology.

We can summarise the main reasons for this pattern of involvement — and the low level of participation in the advance planning of technological change:

- ☐ management resistance to involvement in the phase of planning technological change
- ☐ the complex and diffuse nature of the decision-making process
- ☐ the representational problems that it poses for most enterprise workforces
- ☐ the invisibility and intangibility of the early phases of decision-making.

Why Participation Increased in the Implementation Phase

There is a much greater frequency and intensity of involvement at the implementation phase of technological change. A range of factors promoting participation affects both Management and Employee Representatives.

Management

> **M** In the implementation phase, management resistance to employee representative involvement is minimised by several concerns, in particular:
>
> ☐ Participation provides procedures for solving problems that only become visible when new technologies are implemented.

- ☐ Participation provides a means for regulating changes in the labour supply, in work organisation, and in employment conditions.

- ☐ Participation ensures workforce commitment to and cooperation with change. Management is highly dependent upon workforce cooperation and adaptation when new technologies are being installed, commissioned and operated. Technical problems may arise. Particular demands may be made on the workforce in the period of change-over. Large numbers of workers may be involved in adapting and learning to use the new systems. They may make an important contribution to overcoming "teething troubles" and other problems with the system. It is also recognised that employee responses and attitudes can be critically important in determining the success of the new system.

- ☐ At the same time managers may accept participation in implementation because it does not compromise managerial prerogative, particularly where managers have retained exclusive control over the priorities and objectives of the planning and design of change, which may be "built in" to the designed system.

Employee Representatives

E   As new technologies are installed, technological change becomes highly visible. As new systems are commissioned and employees begin to operate them their employment implications become crystallised. These factors will tend to increase workforce pressure for regulation of the change and of its outcome for employment.

☐   Employee representatives have traditionally been accustomed to influencing substantive issues in relation to change at work, and may therefore prefer to push on issues of substance rather than on the unfamiliar terrain of technological design and planning.

☐   Finally, when new systems are being installed, opportunities arise for "hands on" involvement in change by the workforce and its representatives.

Even though there is an increase in employee representative involvement during the implementation phase, there is still a high proportion of respondents — over 30% overall — reporting no involvement. Many of the reasons advanced to explain lack of involvement in the planning of technological change continue to operate in the implementation phase. However, as the later section will show, the pattern of participation is very uneven between countries. Some national systems with little tradition of employee involvement have a much higher frequency of cases with "no involvement", and much lower levels of intense forms of involvement.

Further light is thrown upon the pattern of participation by the survey evidence regarding the impacts of participation. The attitudinal survey examined how those directly involved in participation evaluated its effects on a range of issues. The impacts of participation were seen, by and large, to have been beneficial.

The Benefits of Participation

Although the overall provision of participation may be unbalanced in terms of its timing and content, that which took place was valued by the respondents. A positive pattern emerges from the attitudinal responses to questions on the effect of participation on a range of enterprise issues.

Respondents were asked whether a range of enterprise issues had got better, got worse or were unchanged by participation. In the next two tables *(Tables 5 and 6)* the bar to the right of the line shows the percentage saying that matters had improved. The left of the line shows the percentage saying that they had got worse. The shorter the block, the more people said that there was no change. In many cases, respondents said that there had been no change. Where participation did have an effect, it was overwhelmingly seen to be positive.

Management assessments of the impacts of participation were positive, and were most marked in relation to workforce cooperation with new technology, workforce attitudes to future technological change, and the utilisation of workforce skills.

Category	Got Worse %	Got Better %
QUALITY OF DECISION-MAKING	-2	30
TIME NEEDED FOR DECISION-MAKING	-8	23
TIME NEEDED FOR IMPLEMENTATION	-1	28
ATTENTION TO WORKFORCE CONCERNS	-5	39
UNDERSTANDING OF MANAGEMENT CONCERNS	-5	38
INDUSTRIAL RELATIONS	-4	25
IDENTIFICATION WITH COMPANY	-5	30
COOPERATION WITH NEW TECHNOLOGY	-3	60
ATTITUDES TO FUTURE TECHNOLOGY INTRODUCTION	-5	52
UTILISATION OF SKILLS	-2	54

ASSESSMENT OF PARTICIPATION

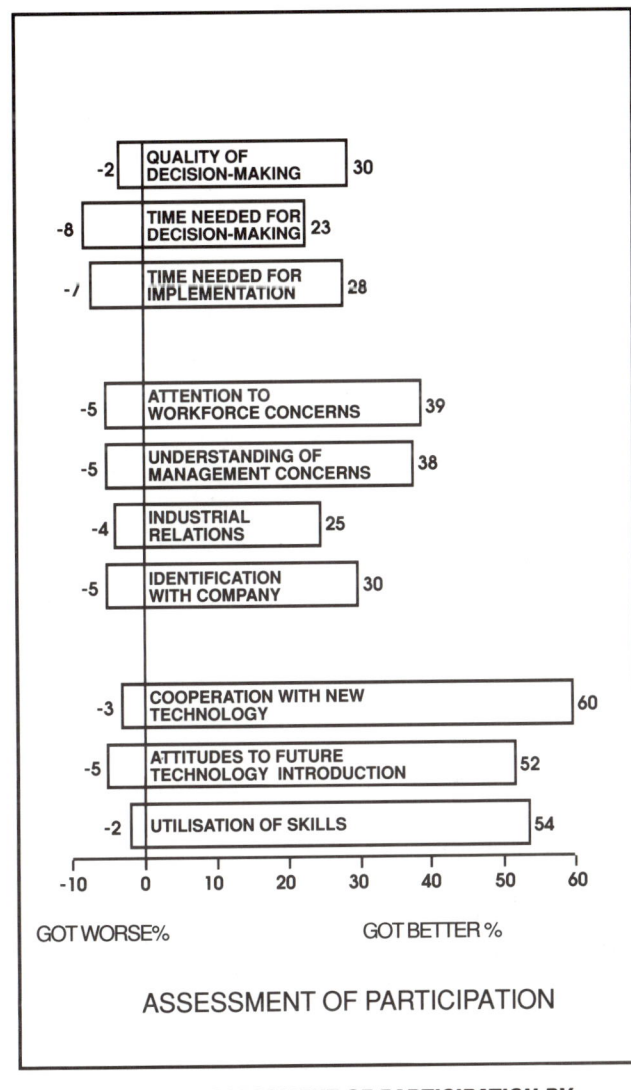

TABLE 5: ASSESSMENT OF PARTICIPATION BY EMPLOYEE REPRESENTATIVES

Category	Got Worse %	Got Better %
QUALITY OF DECISION-MAKING	-1	33
TIME NEEDED FOR DECISION-MAKING	-10	21
TIME NEEDED FOR IMPLEMENTATION	-8	26
ATTENTION TO WORKFORCE CONCERNS	-1	49
UNDERSTANDING OF MANAGEMENT CONCERNS	-2	46
INDUSTRIAL RELATIONS	-1	27
IDENTIFICATION WITH COMPANY	-2	34
COOPERATION WITH NEW TECHNOLOGY	-2	61
ATTITUDES TO FUTURE TECHNOLOGY INTRODUCTION	-3	54
UTILISATION OF SKILLS	-1	52

ASSESSMENT OF PARTICIPATION

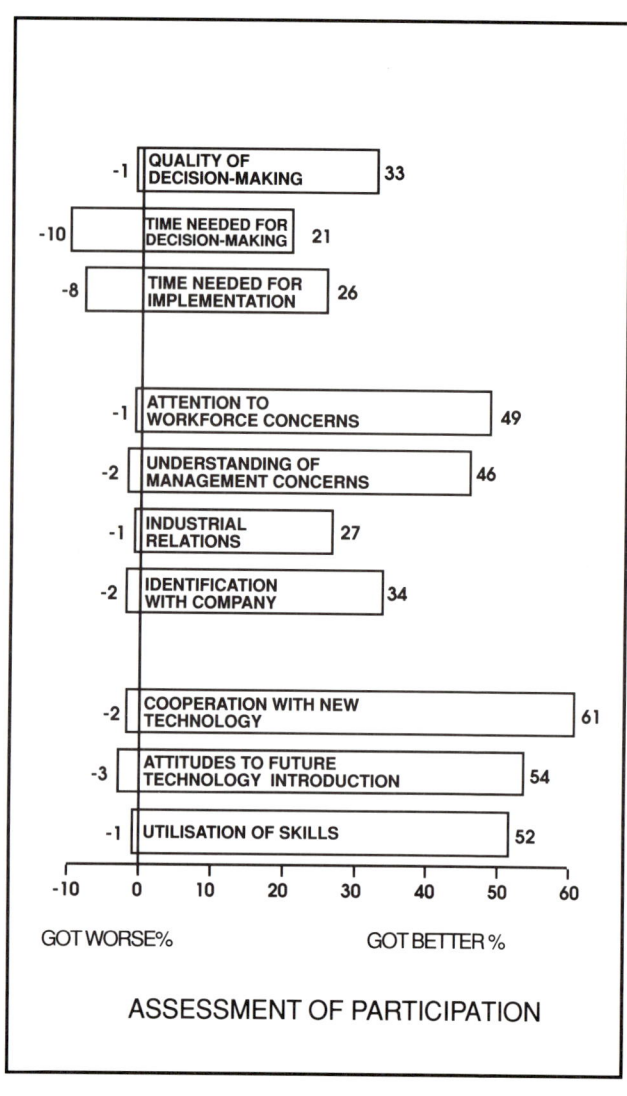

TABLE 6: ASSESSMENT OF PARTICIPATION BY MANAGEMENT

Managers also reported that participation led to greater management attention to workforce concerns, and to improved understanding by the workforce of management concerns.

Negative impacts were most apparent, and the fewest positive impacts of participation were noted, in relation to the time needed for decision-making and implementation of change. Even here, positive assessments still outweighed negative assessments by over 2 to 1 — and most people (over two-thirds of the total) recorded no change.

Employee representatives' evaluations show a very similar picture, with workforce cooperation, attitudes to technology and the utilisation of skills highlighted as having greatly improved. Employee representatives were significantly less enthusiastic than managers about the positive effects of participation, either on management attention to workforce concerns or on workforce understanding of management concerns (recorded as improving in 39% and 38% of cases by employee representatives, compared to 49% and 46% respectively by managers). However, employee assessments remain overwhelmingly positive. These results confirm that, over a range of issues, both parties saw themselves as having benefitted from involvement. Participation was perceived, at least in part, as a positive-sum game.

Despite the constrained nature of participation and management's neutralisation of it in the early phases, it appears there were sufficient benefits flowing to both management and employees to keep them positive about involvement. This is especially apparent if we look at their attitudes to participation in future technological change.

Involvement in the Future

An important objective of the attitudinal survey was to determine the aspirations of the parties to participation in the future. The greatest differences between management and employee responses were noted here.

Table 7 shows the levels of involvement (all forms) that managers sought in future planning and implementing of technology. It also shows, for comparative purposes, the levels of involvement currently reported. Most managers wanted to see an increase in some form of involvement, particularly in relation to involvement in the advance planning stages; the increase was of the order of 21 points. 9 out of 10 managers want some form of involvement at the implementation stage and 4 out of 5 want involvement at the planning stage.

> ❛ Despite the constrained nature of participation and management's neutralisation of it in the early phases, there still appeared to be sufficient benefits flowing to both management and employees to keep them positive about involvement. ❜

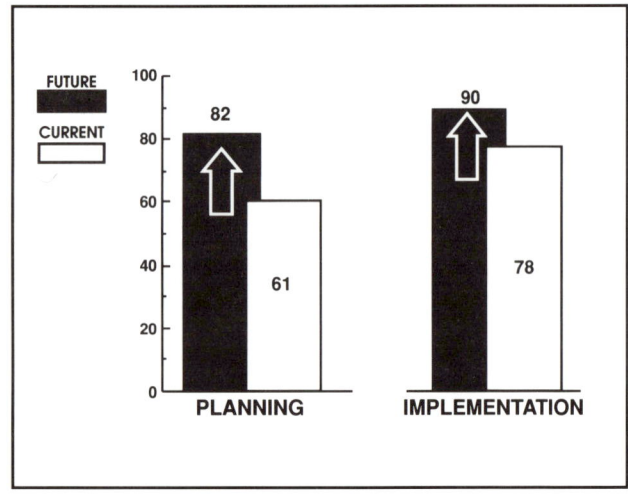

	PLANNING	IMPLEMENTATION
FUTURE	82	90
CURRENT	61	78

TABLE 7: INVOLVEMENT CURRENTLY AND SOUGHT IN FUTURE BY MANAGEMENT

Table 8 shows the responses of employee representatives. The increase sought in levels of future involvement was even more dramatic amongst employee representatives — from 57% currently, to 91% seeking future involvement. Almost all employee representatives sought some form of involvement in future — 19 out of 20.

The differences between managers and employee representatives become even more marked over the more intense forms of involvement (through negotiation or joint decision-making).

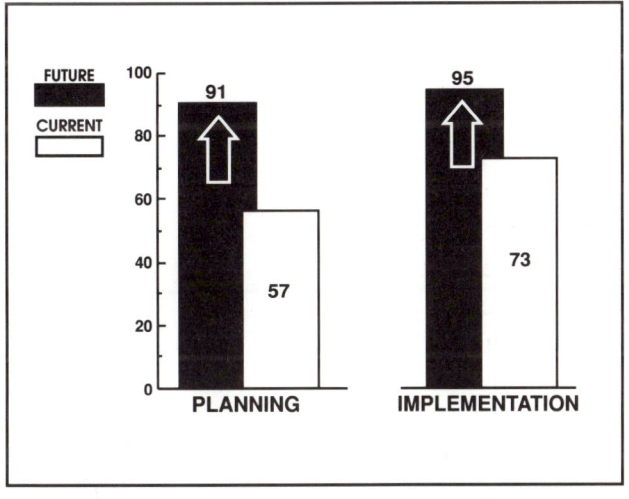

TABLE 8: INVOLVEMENT CURRENTLY AND SOUGHT IN FUTURE BY EMPLOYEE REPRESENTATIVES

As *Table* 9 shows, more intense forms of involvement are currently found in only a small proportion of cases (only 10% in planning and 18% in implementation of new technology). A larger number of managers favour more intense forms of involvement in future — half as many again as the current levels. Thus 17% of managers want intense involvement in the planning and 29% in the

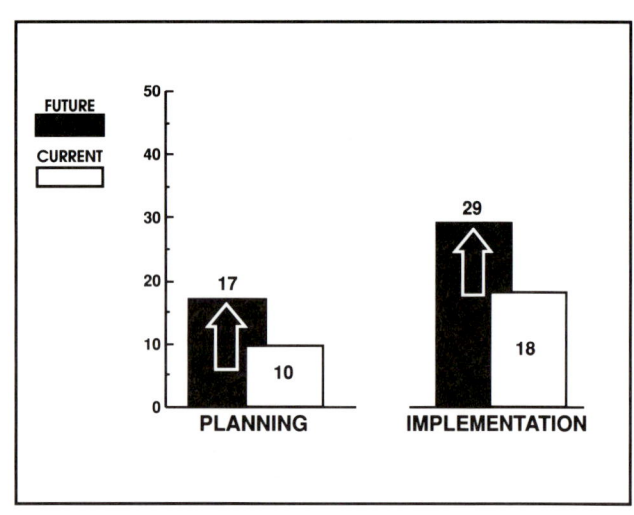

[Bar chart showing FUTURE (black) and CURRENT (white) bars. Planning: 17 future, 10 current. Implementation: 29 future, 18 current.]

TABLE 9: MORE INTENSE FORMS OF INVOLVEMENT CURRENTLY, AND SOUGHT IN FUTURE BY MANAGEMENT

implementation stage. Support for more intense forms of involvement in future is strongly correlated with the existence of intense involvement at present. It is also correlated with a positive evaluation of current participation. This increase in the proportion of managers seeking more intense involvement is important. However, intense involvement is still rare, so the level of managers seeking such involvement in future remains relatively low. Only a minority of managers wish to see intense involvement in future.

Employee representative responses show a very different pattern, as summarised in *Table 10*. 47% wanted to see more intense involvement in the implementation stage. This represents a three-fold increase over the current level of intense involvement. In relation to planning, the increase was even more dramatic. The proportion of

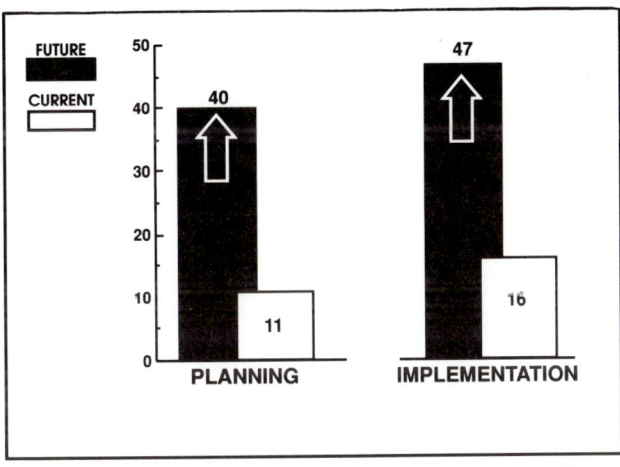

TABLE 10: MORE INTENSE FORMS OF INVOLVEMENT CURRENTLY, AND SOUGHT IN FUTURE BY EMPLOYEE REPRESENTATIVE

employee representatives seeking more intense involvement in future planning almost quadrupled over current levels, to 40%.

Workforce representatives sought greater intensity and earlier involvement in the introduction of technological change. When we look at the content of participation below, we find that they sought involvement in virtually all aspects of decision-making. In contrast to management responses, support for high levels and intensity of involvement was not confined to those cases where a high intensity of involvement currently existed, or where employee representatives gave a positive evaluation to current involvement. Employee representatives sought an extension of involvement across the board. However, employee representatives' ambitions were guided by current achievements; employee representatives' aspirations were highest in those countries in which early and intense forms of involvement already existed.

		All forms	More intense forms
Primary workforce concerns	HEALTH AND SAFETY	93	47
	JOB SATISFACTION	91	39
	TASK SPECIFICATION	93	39
	WORK ORGANISATION	93	37
Primary management concerns	PRODUCTION FLEXIBILITY	85	20
	REDUCING COSTS	81	25
	INVESTMENT CRITERIA	68	14
	MARKET STRATEGY	67	12

ALL FORMS OF INVOLVEMENT

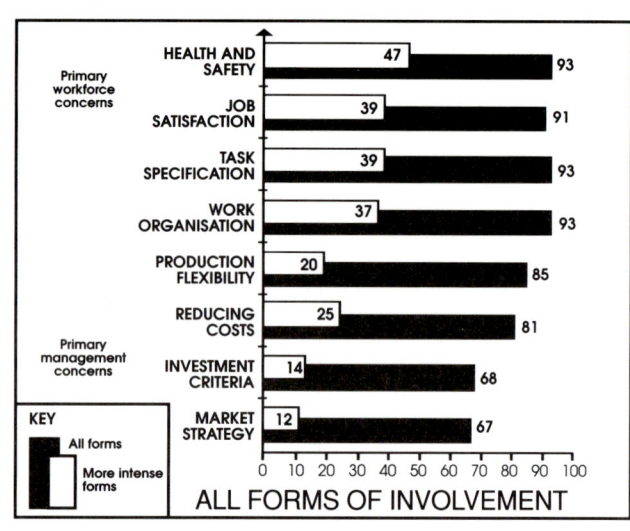

TABLE 11: COVERAGE OF FUTURE INVOLVEMENT BY MANAGEMENT

Table 11 shows the frequency with which managers sought involvement in future over a range of issues. It shows levels of future coverage sought both for intense forms of participation and for all forms of participation. When we compare this figure with *Tables 3* and *4*, which looked at the coverage of these issues in current involvement (respectively all forms and intense forms), we find that the same overall pattern is apparent. The coverage of future involvement continues to be weighted towards issues on the right hand side of the graph — namely employment issues such as health and safety, task specification and work organisation. Significantly fewer managers seek the involvement of employee representatives in those issues which can be seen as primary management concerns, particularly in relation to long-term management strategy (eg for markets or investment criteria). The coverage sought by managers over operational management concerns — such as cost reduction and the introduction of new products — falls mid-way between the two. So the traditional lines of managerial prerogative are still apparent. These are much more strongly marked if we refer to more intense forms of involvement.

However, it is also clear that managers supported extensions of future involvement that implied further erosion of the traditional lines of managerial prerogative. Thus managers sought higher levels of involvement in future across the range of issues (the overall increase is around 10% for all forms of involvement) and particularly over strategic management concerns. (The increase in future involvement sought by managers, over market strategy or investment criteria, is over 20% for all forms of involvement). Two-thirds of the managers surveyed were willing to countenance some involvement of employee representatives in primary strategic concerns. A similar development is apparent when we look solely at intense forms of participation. Although intense forms of involvement are currently low, twice as many managers as currently extend it wanted such involvement in future.

Table 12 shows the same data for employee representatives. The same general pattern is visible, with involvement weighted towards joint concerns of management and workforce alike, though here the pattern is much less marked. Most employee representatives wanted to see some form of involvement on the whole range of issues. They were even keener than managers for current involvement to be extended in future (an overall increase of around 20%), and particularly in

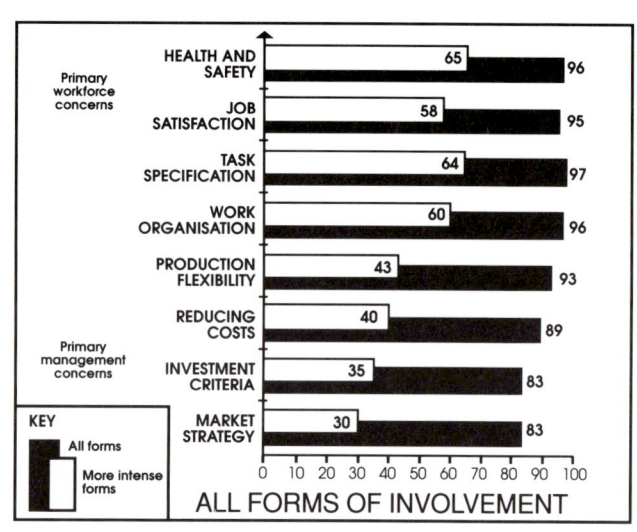

TABLE 12: COVERAGE OF FUTURE INVOLVEMENT BY EMPLOYEE REPRESENTATIVES

Category		All forms	More intense forms
Primary workforce concerns	HEALTH AND SAFETY	65	96
	JOB SATISFACTION	58	95
	TASK SPECIFICATION	64	97
	WORK ORGANISATION	60	96
Primary management concerns	PRODUCTION FLEXIBILITY	43	93
	REDUCING COSTS	40	89
	INVESTMENT CRITERIA	35	83
	MARKET STRATEGY	30	83

> ❛ The greatest differences between management and employee representative responses were found in relation to aspirations towards participation in future. ❜

relation to management strategic issues (where a 40% increase was sought). Thus five out of six employee representatives wanted some form of involvement in market strategy and investment criteria, and nineteen out of twenty wanted involvement in "joint concerns" such as work organisation and health and safety. Employee representatives were particularly keen to see an extension of more intensive forms of involvement in future. They wanted intense involvement over "joint concerns" in twice as many cases as currently found. They also wanted a five-fold increase in intense involvement over management strategic concerns. However, because intense involvement in the latter is currently very low, the levels of intense involvement sought by employee representatives continue to be weighted towards the "joint concerns" of management and workforce.

Employee representatives are not reluctant to become involved in future in areas traditionally seen as managerial prerogative. Indeed, they are specially keen for such involvement to be more frequent, and more intense. Managers do still exhibit a continued attachment to the traditional boundaries of managerial prerogative, but the majority are willing to see some form of involvement in these strategic issues.

The future aspirations of the social partners

A major objective of the attitudinal survey was to determine the future aspirations of both managers and employee representatives with regard to participation, and thus to uncover the scope and value for improving participatory arrangements. A large amount of data have been produced and are being analysed. It is not possible to present all this information in this short report. The overall findings about management and representative responses can be briefly summarised here.

In most cases managers sought a continuation of existing arrangements. However, a significant minority sought some extension of involvement. Managers' support for the continuation and the extension of involvement is strongly related to the extent to which they currently involve employee representatives in technological change, and to their positive evaluation of this involvement. Management aspirations for the future are thus pragmatic and guided by current experience.

The greatest differences between management and employee representative responses were found in relation to aspirations towards participation in future. Employee representatives in almost all cases sought an extension of involvement, and an increase in the intensity of involvement. Employee representative ambitions for the future were "realistic" insofar as they were constrained by the extent to which early and intense forms of involvement had already been achieved. However, employee representatives' aspirations were not determined by their evaluation of current participation. Arguably this is because representatives who were critical of current participation were likely to call for an extension of involvement (to overcome the shortcomings of current involvement), whilst those with positive current experience were keen to maintain and extend involvement. This position is not surprising in that it corresponds to the respondents' role as employee representatives.

CHAPTER 4
INTERNATIONAL COMPARISON

Different National Models

There is a great variety of participatory practices across the European Community. There are important differences between the twelve member states in their industrial relations systems; their institutional structures, their traditions, and their broader legal and policy contexts. The variety of forms of involvement, whilst influenced strongly by the domestic system of industrial relations, was not determined by that system. There are differences between countries in terms of their participative provisions, but there is also a range of different practices within the countries studied. Any comparative study should bear such caveats in mind when attempting to generalise from the evidence of case studies and of survey material.

In the Foundation's earlier *Participation Review*, four systematic approaches to participation were put forward. One was based upon the statutory or legalistic model found, for instance, in Germany and The Netherlands. A second was based on the Danish model of centralised agreement between the major social actors. The voluntaristic approach exemplified by the British, and to some extent the Italians, was also used. Finally, the Review identified an "organic" or paternalistic approach which emulated Japanese participative practice. Subsequently, the Commission's discussions of participatory models in the context of the European Company Statute have adopted a similar differentiation of approaches, replacing the "organic model" with one akin to the enterprise "works council" arrangements found in France and in Italy. These approaches or models may be useful to us in assessing the results of the twelve country study. First we examine their relevance to characterising the range of national situations. We then explore how far they can explain the different national patterns of participation.

In order to show the main pattern of results we will simplify the discussion and presentation of data by focussing on six of the countries that represent exemplars of different approaches. (The data for the other six European nations are included as an appendix to this report.)

The German case, because of its statutory backing, stands as one important focus; so too does the Danish case with its mature set of centralised institutions and agreements. The Belgian case, with its mix of legal and bargaining provisions, represents a distinct intermediate position between the first two. The UK is a clear model for the voluntaristic approach. Italy represents the factory works council approach identified by the Commission. The current system of industrial relations in Spain has been developed rapidly over the recent period. All of these six countries have well-developed systems of industrial relations and they exhibit significant differences in the degree of formality of industrial relations, in their use of legislation in their institutional structure and their level of operation (eg between plant, industry and national levels).

Given these broad categorisations, what does the comparative research reveal in the way of results and patterns of national responses? *Tables 13* and *14* compare the extent of all forms of participation across the six European nations. These show the responses of management and employee representatives regarding current involvement in planning and implementing technological change, and their wishes for such involvement in future. *Tables 15* and *16* show the same data for more intense forms of involvement.

First we shall examine the current levels of involvement in the different countries. The first

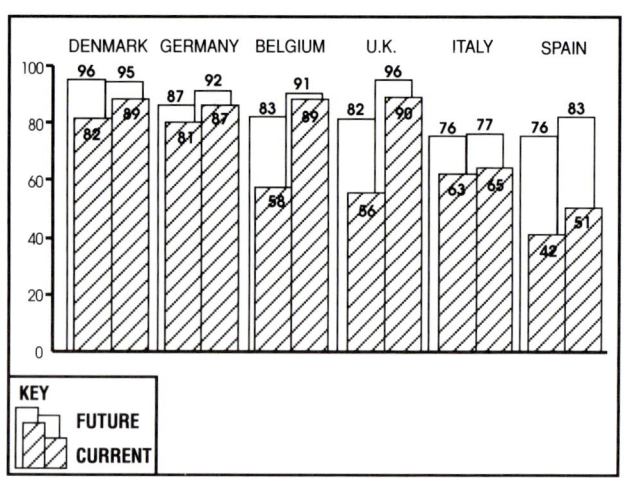

TABLE 13: INTERNATIONAL COMPARISON
(All forms of participation) Current & Future Participation by Management

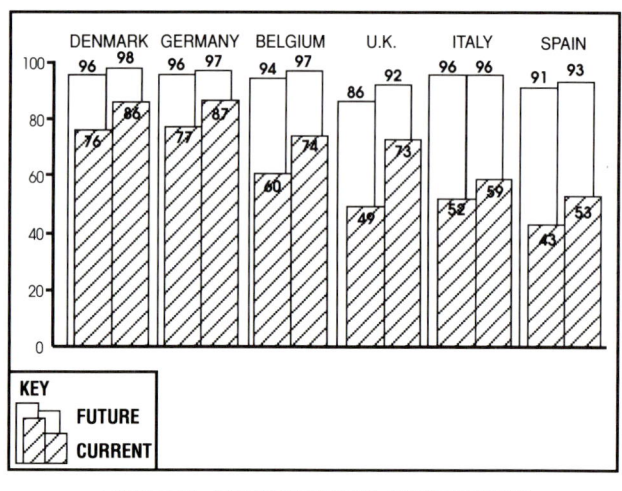

	Denmark	Germany	Belgium	U.K.	Italy	Spain
Current	82	81	58	56	63	42
Future	96 / 95 / 89	87 / 92 / 87	83 / 91 / 89	82 / 96 / 98	76 / 77 / 65	76 / 83 / 51

TABLE 14: INTERNATIONAL COMPARISON
(all forms) Current & Future Participation by Employee Representatives

	Denmark	Germany	Belgium	U.K.	Italy	Spain
Current	76 / 86	77 / 87	60 / 74	49 / 73	52 / 59	43 / 53
Future	96 / 98	96 / 97	94 / 97	86 / 92	96 / 96	91 / 93

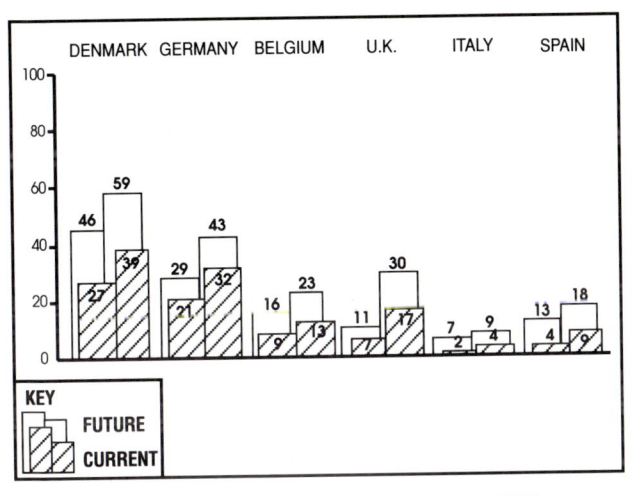

TABLE 15: INTERNATIONAL COMPARISON
(more intense forms) Current & Future by Management

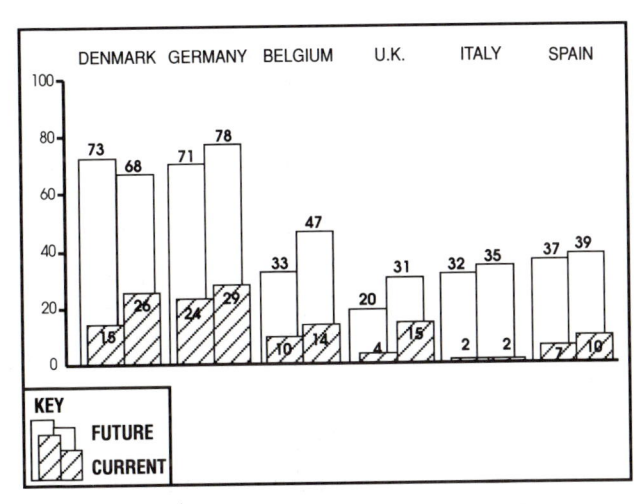

	DENMARK	GERMANY	BELGIUM	U.K.	ITALY	SPAIN
Current	27	21	9	11	2	4
Future	46 / 59 (39)	29 / 43 (32)	16 / 23 (13)	17 / 30	7 / 9	13 / 18 (9)

TABLE 16: INTERNATIONAL COMPARISON
(more intense forms) Current & Future Participation by Employee Representatives

	DENMARK	GERMANY	BELGIUM	U.K.	ITALY	SPAIN
Current	18	24	10	4	2	7
Future	73 / 68 (26)	71 / 78 (29)	33 / 47 (14)	20 / 31 (15)	32 / 35 (2)	37 / 39 (10)

67

point to note is the broad overall agreement between management and employee representative responses regarding the current extent of involvement; only the UK and Belgium exhibit a measure of difference here.

There were, however, significant differences in the levels of current involvement. *Table 13* shows management responses for current participation of all forms. Germany and Denmark, at one extreme, are countries with a high level of involvement in the advance planning stage as well as in the implementation stage. Involvement is initiated early in decision-making. These are perhaps features of highly formalised participation procedures, derived from legislation or from central contractual agreement. Italy shows much lower levels at both stages. Belgium and the UK show an interesting intermediate position, with much lower levels of involvement at the planning stage than at the implementation stage. In Britain, this is perhaps a hallmark of a largely reactive system of industrial relations, and can be attributed to the voluntaristic basis of collective bargaining and the decentralised focus of union representation. This often operates at the level of the work group rather than the enterprise or sectoral level. Despite these differences there are also common features. In particular, the current levels of participation are consistently higher in the implementation phase than in the planning phase in *all* countries.

When we look at the figures for more intense forms of participation, shown in *Tables 15* and *16*, the situation is distinctly different. Across the board, the current levels of intense forms of participation are much lower and in some countries this form of participation is almost absent — this is especially

> *In all countries, the number of managers seeking more intense forms of future involvement increased significantly over the levels reporting such involvement currently*

true of Italy and Spain, where figures below 10% were found.

The differentiation between countries is much greater for intense forms than for all forms of participation. As far as current involvement in the implementation phases is concerned, there are greater differences between Italy and Denmark in the frequency of more intense forms of involvement (4% as opposed to 39%) than there are for all forms of involvement (66% as opposed to 89%). The low levels of intense participation in Italy may partly be related to a distinctive feature of the Italian context. In Italy, a main avenue for involvement over technology has been through collective agreements on information disclosure. This may explain why a large proportion of Italian respondents reported that participation took the form of the provision of information. Overall it is clear that national industrial relations systems with high levels of current participation of all forms had disproportionately higher levels of more intense forms of participation. Only Germany and Denmark have significant levels of intense participation currently, with a third of managers reporting negotiation or joint decision-making over the implementation of change.

Respondents' aspirations towards participation in future show some very striking features. Across all of the different national systems, there is a desire for higher levels of participation than currently exist. A high proportion of managers as well as employee representatives sought some form of future involvement. The figures for employee representatives in particular reflect a very high demand for participation in the future. Except in the case of the UK, over 90% of employee respondents sought some form of involvement in the advanced planning stage. Italy had the least enthusiastic managers, but again figures of 76% and 77% cannot be seen as signalling a critical or negative stance.

Another point to note is that the desire for increased participation in future was greater in relation to involvement in advanced planning than in relation to implementation. The levels of involvement sought over planning were near or equal to the levels of involvement sought in implementation. In other words, management and employee representatives alike wanted earlier involvement, as well as a general increase in involvement, and they wanted this involvement to start from the earliest stages of planning technological change.

The picture regarding future aspirations towards participation gets much more complex for more intense forms of involvement. *Table 15* summarises management responses.

In all countries, the number of managers seeking more intense forms of future involvement increased significantly over the levels reporting such involvement currently. In those countries in which intense forms of involvement were currently common, managers were likely to want even greater levels of intense involvement in future. Again, Denmark emerges as the country where management sought the highest levels of future intense involvement (59%, as opposed to 9% of Italian managers) in relation to implementation. Finally, managers were less likely to want intense involvement in planning than in implementation. This contrasts with the picture we found for all forms of involvement where support for future involvement was almost the same for planning as for implementation.

The comparison of future wishes reveals differences between the actors; whereas management in general sought modest increases from the existing pattern, the employees sought more substantial increases. This was true for all countries — though the pattern differs slightly from what might be expected from our different models of industrial relations, and from the earlier finding that expectations for future involvement were in proportion to the extent of involvement currently achieved. As expected, the German and Danish employee representatives were the most keen for intense involvement in future. Italian employee representatives were marked by their enthusiasm for higher levels of intense involvement (which increased from 2% currently to 32%, in relation to planning, and 35% over implementation). Spain showed similarly high ambitions for the future. The UK, in contrast, was marked by the relatively limited aspirations of employee representatives for intense involvement in future — the lowest of the six countries. Such low ambitions by British employee representatives may in part reflect a degree of demoralisation after an extended period which has been economically and politically unfavourable to trade union and workforce bargaining power.

A particularly notable aspect is the emphasis given by employee representatives to increasing the frequency and intensity of advance involvement — in the planning stage. Indeed Danish employee representatives were more interested in securing intense involvement at the planning stage than at the implementation stage.

The picture in the other six European countries follows a broadly similar pattern, with very high levels of support for future participation of some form. The current extent of intense forms of participation is limited — only Ireland reports over 20% in this category, whilst Portugal and Luxembourg show similarities with the Spanish and Italian situations, exhibiting figures below the 5% mark. Support for more intense future involvement exists across the board, and is particularly marked in France, Greece, Ireland and Holland.

These results tend to confirm the relevance of the broad models or approaches to participation. The pattern of current participation in technological change can readily be related to the characteristics of the national industrial relations systems. They also assist in interpreting respondents' aspirations for participation in future. Thus the existence of legislation or agreed procedures at the central level for participation was associated with greater frequency and intensity of current involvement over technological change, and with early involvement. These provided a context which appears to promote calls for further extension of participation. It is certainly not the case that experience of participation leads to a dampening of the social partners' demands for involvement.

Implications of International Comparison

Whilst there were some important differences between countries, involvement in technological change was not wholly the prerogative of countries with extensive statutory participation rights. Technological change created a series of concrete problems, uncertainties and issues around which participation was considered necessary by managers as well as employee representatives. In this instance participation was indeed seen as a productive force for change. It stemmed from the concrete needs of the parties. It was motivated by the overlapping concerns and interests of managers and employee representatives. It was thus pursued for pragmatic reasons rather than being a product of a set of programmatic legal rights determined outside the enterprise.

The international comparison did, however, show a correlation between the existence of joint decision-making and the existence of central laws and binding agreements. This highlights the fact that there are two sets of factors at work, ie broad contextual factors, and the particular enterprise factors that determine the actual form and extent of participation. The attitudinal survey and the results of the case-studies highlighted these different factors favouring participation.

Factors Favouring Participation

Contextual

- ☐ Legal Requirements Promoting Participation
- ☐ Institutional Arrangements at National or Sectoral Level
- ☐ Maturity of the Industrial Relations System
- ☐ Supportive Labour Market Policies (eg for retraining or employment security)

Enterprise

- ☐ The Tradition of Participation (on technology and other issues)
- ☐ Management Dependence on Workforce Skills
- ☐ The Complexity of Change and the Technical and Organisational Uncertainties Involved

- ☐ Where Labour Reduction is not the Primary Aim
- ☐ Incremental rather than Radical Innovation
- ☐ High Proportion of Professional Workers.

This illustrates two very different sets of influences. The first is contextual, and will be generalised at the national or industry level and have a relatively stable and long-term effect, for instance the operation of the German co-determination laws. The other will tend to vary between enterprises depending upon the characteristics of production processes, the technological change occurring, and the sets of detailed factors particular to the enterprises — their occupational structures, their traditions of participation and their participation strategies.

The participation that does exist develops in response to a host of structural pressures and to the differences in the strategies of the social partners. Taking these factors together, it is little wonder that one finds such variation in participation provision both within countries and across countries.

CHAPTER 5
CONCLUSIONS

Conclusions

We have examined the overall pattern of participation and the way it is weighted in favour of operational issues found in the implementation phase. Our analysis suggests that the resistance of managers to more intense forms of participation, and the relative exclusion of employee representatives from involvement in the early stages and strategic aspects of decision-making, may be indicative of a management strategy of neutralising employee involvement. In other words, managers engage in participation with employee representatives (and may gain benefits from so doing), but in a manner which limits the ability of employee representatives to influence the decision-making process.

This highlights the issues raised in the introduction about what model of participation is uppermost and being realised in practice. Is it participation as a productive force, as a democratic force or as a positive-sum game? From the evidence, it might appear that managers' concerns were being dealt with more effectively than those of the workforce. The existing pattern of involvement inside enterprises seems to be favouring participation as an agent for efficiency rather than an agent for redistributing power.

Though there was evidence that participation was fulfilling the "secondary concerns" of employee representatives (for example, the perceived benefit of participation from management paying greater attention to workforce concerns), these were being achieved as an *indirect* consequence of participation, rather than through *direct* influence by employee representatives over the decision-making process. This unbalanced picture calls into question whether participation is operating as a positive-sum game for both parties if the primary aims of only one of the parties are being met.

Questions for the Social Partners

> The main problems for employee representatives arose from the serious deficiencies of access to the strategic sites of decision-making in the early planning phases.

By way of conclusion, we do not wish to repeat the evidence shown earlier. Instead, we raise the questions that the findings pose for the major actors.

Questions for Workforce Representatives

The evidence from both phases of research shows workforce representatives to have been open and responsive to technological change, and certainly not the Luddites described in earlier debates. Further, they had high ambitions for participation in future.

The main problems for employee representatives arose from the serious deficiencies of access to the strategic sites of decision-making in the early planning phases. Here the influence of workforce involvement on managerial decision-making was marginal, though the case studies revealed that there may be important informal channels of representation. The question remains as to whether employee involvement of this kind can have important impacts on the overall process of technological change.

Participation had important indirect impacts, however. The importance that management attached to winning workforce consent to change through participation forced management to pay attention to the concerns and responses of the workforce. Employee concerns were represented by proxy, accommodated through managerial interests rather than through the direct representation of employees. In general, employee representatives were keen to see more participation. However, this raises a number of questions for employee representatives:

1. *Is participation that meets their secondary concerns enough, or does it only offer a*

situation where they accept a second best form of involvement — especially if it offers them responsibility without any real power?

2. *Despite these limitations, the case studies revealed opportunities for workforce influence where the representatives had concrete, forward-looking strategies. How far have employees and their organisations developed such strategies? Have clear strategies been developed on the part of the workforce, and do the workforce and their trade unions have the technical resources and expertise to underwrite or create such strategies?*

3. *What practical and ideological problems present themselves in the development of trade union strategies? Do employee representatives have appropriate structures to develop such strategies and to articulate them at the best levels within the organisation and beyond?*

4. *The research raises important questions regarding whether employee representatives will continue to support involvement if it does not yield real opportunities for influence. At what point should workers and their representatives withdraw support for participation?*

Questions for Management

Managers were more divided than employee representatives; their positions were sometimes contradictory and ambiguous. This seems to reflect the dilemmas they face. They are keen to see participation, but of a particular kind with defined boundaries, that does not, for instance, challenge their prerogative in the key strategic and business areas. Managers have been most willing to engage in

> ❛ They (management) are keen to see participation, but of a particular kind… ❜

participation in the final stages of technological change — when it is being implemented — and are largely resistant to more intense forms of involvement. The character of participation therefore appears to be one that involves the workforce in the problems of running the enterprise without giving commensurate power to jointly resolve the problems raised. Management's willingness to countenance the continuation and extension of participation derives from the benefits they currently receive from involvement initiatives.

Questions raised for managers are:

1. *The current European discussions around participation offer management moves towards the extension of involvement. How should participation that meets management wishes be developed?*

2. *If managers are the gatekeeper to participation and want more participation, why do they not initiate this within the enterprise on a voluntary basis? Are the main problems of a practical nature, or do ideology and the protection of their elite position figure in the lack of extensive involvement?*

3. *We see that enterprises thrive, where participation is strong, at least as well as those where there is little involvement. Participation clearly does not impair the success of the organisation. Therefore, why do many managers resist employee involvement? Why does management seek to preserve its prerogatives?*

4. *What are the key fears of management, and the obstacles to removing some of the boundaries they set regarding the limits of involvement?*

5. *What is the optimal trade-off between the dilution of the monopoly of managerial prerogative and control over decisions, and gaining higher levels of workforce commitment, especially trade union commitment, to technological change?*

6. *How would management respond if trade unions adopted a more critical strategy and threatened to withdraw from participation structures that offered them only marginal benefits? Would managers be willing to give employee representatives more influence over technological decisions?*

7. *Is management more interested in developing direct forms of participation with individual employees, sidestepping collective participation with employee representatives, with all the implications of this procedure?*

Questions for Policy-makers

The current period is one of reassessment regarding participation and the correct roles for the social partners at the European level. The discussions around the Social Dialogue, Community Charter of Basic Social Rights for Workers and the European Company Statute all involve the issues under discussion here. The Commission has a direct and practical interest, as it has the responsibility for drafting frameworks and legislation in this area and for persuading the social actors that such initiatives should be accepted.

1. *Does the Commission's policy for employee involvement have a clear basis in terms of the models alluded to in the research? For example, does it derive in the main from a*

concern with efficiency (participation as a productive force), or from a concern with the redistribution of power and resources (participation as a democratic force), or from the expectation that participation can satisfy the mutual interests of both parties?

2. *We note the variety of instruments under consideration by policy-makers at the European level. These range from a voluntary approach, to bi-partite and tri-partite initiatives, and to legislative approaches. What models of participation underpin these different models? How do the findings of the Foundation survey bear upon the choice between these different instruments for intervention?*

3. *In particular, it appears that the voluntary recommendations of the Social Dialogue Joint Opinion on greater consultation during the introduction of new technology are not being met in practice. Why are they not being fulfilled? What are the implications for the European Commission?*

4. *The survey highlights the wide variation between individual European nations in terms of the frequency and intensity of participation. This reflects differences in the tradition, structure and strategies of the social partners and their varying frameworks of national legislation. Does this mean that participatory practices across Europe will remain patchy and uneven? Can voluntary measures deal with this?*

5. *Can the Commission design participative initiatives that could promote the early involvement of both sides in the planning and design phase?*

APPENDIX 1

INTERNATIONAL COMPARATIVE DATA ON REMAINING SIX COUNTRIES:
France, Greece, Ireland, Luxembourg, the Netherlands, Portugal.

Table 17: Remaining Six Countries — Current & Future Participation By Management. (Page 83)

Table 18: Remaining Six Countries — Current & Future Participation By Employee Representatives (Page 83).

Table 19: Remaining Six Countries — More intense forms of Participation: Current and Future By Management (Page 84).

Table 20: Remaining Six Countries — More intense forms of Participation: Current and Future By Employee Representatives (Page 84).

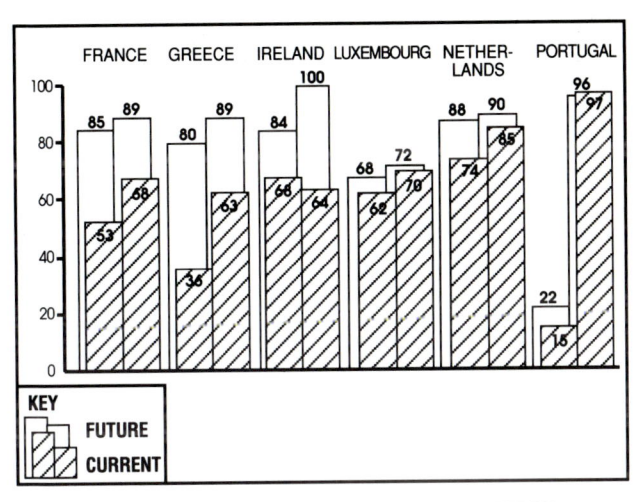

TABLE 17: PLANNING & IMPLEMENTATION
Current & Future (All forms) by Management

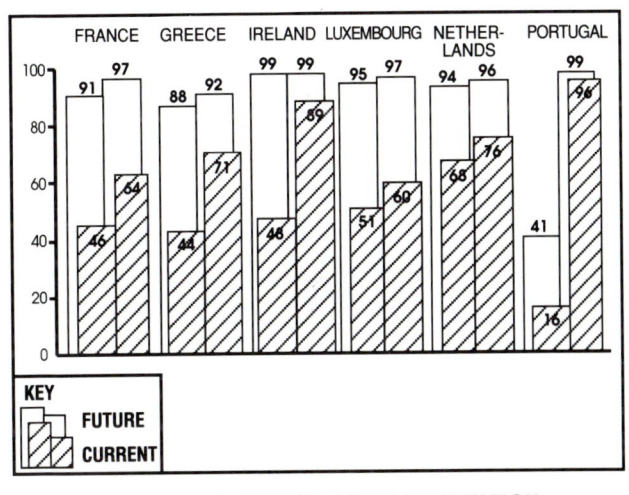

France: 85 / 53, 89 / 68
Greece: 80 / 36, 89 / 63
Ireland: 84 / 66, 100 / 64
Luxembourg: 68 / 62, 72 / 70
Netherlands: 88 / 74, 90 / 85
Portugal: 22 / 15, 96 / 97

TABLE 18: PLANNING & IMPLEMENTATION
Current & Future (All forms) by Employee Representatives

France: 91 / 46, 97 / 64
Greece: 88 / 44, 92 / 71
Ireland: 99 / 46, 99 / 89
Luxembourg: 95 / 51, 97 / 60
Netherlands: 94 / 68, 96 / 76
Portugal: 41 / 16, 99 / 96

83

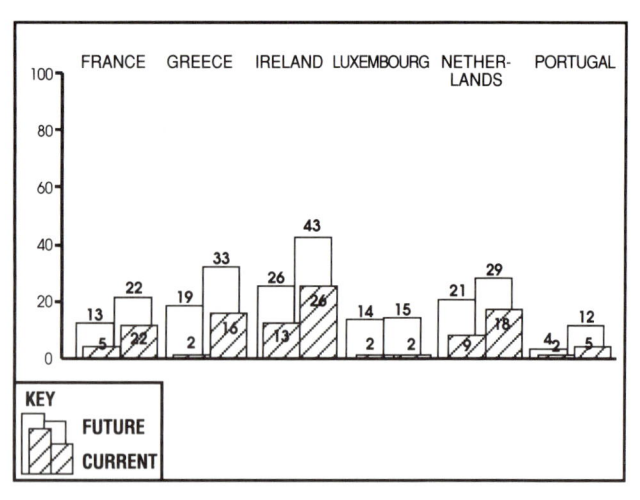

TABLE 19: PLANNING & IMPLEMENTATION
Current & Future (more intense forms) by Management

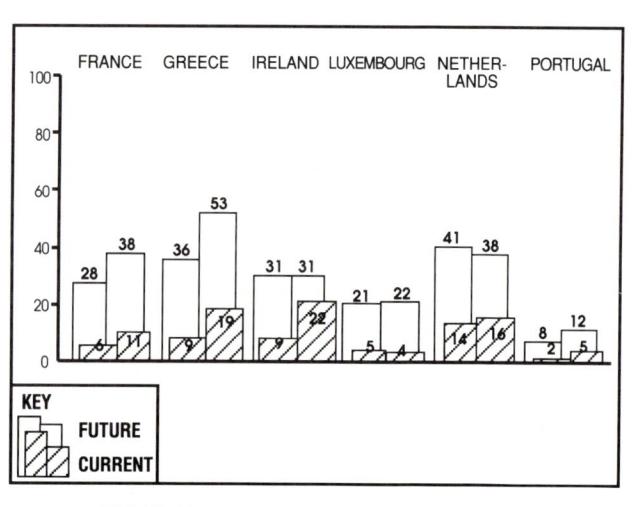

Country	Current	Future
FRANCE	5	13
FRANCE	22	22
GREECE	2	19
GREECE	16	33
IRELAND	13	26
IRELAND	26	43
LUXEMBOURG	2	14
LUXEMBOURG	2	15
NETHERLANDS	9	21
NETHERLANDS	18	29
PORTUGAL	2	4
PORTUGAL	5	12

TABLE 20: PLANNING & IMPLEMENTATION
Current & Future (more intense forms) by Employee Representatives

Country	Current	Future
FRANCE	6	28
FRANCE	11	38
GREECE	9	36
GREECE	19	53
IRELAND	9	31
IRELAND	22	31
LUXEMBOURG	5	21
LUXEMBOURG	4	22
NETHERLANDS	14	41
NETHERLANDS	16	38
PORTUGAL	2	8
PORTUGAL	5	12

APPENDIX 2

Relevant European Research Studies

— **European Foundation for the Improvement of Living and Working Conditions, Dublin:** *"The role of the parties concerned in the introduction of New Technologies. Phase 1"* Consolidated Report by P. Cressey), 1984

— European Foundation for the Improvement of Living and Working Conditions, Dublin: *"The role of the parties concerned in the planning and design of new forms of work organisation"* (Consolidated Report by G. Della Rocca), 1983

— **Roskin College:** *"Workers and New Technology; disclosure and use of company information"*, 1984

— **CEE-DGV:** *"Etude sur les liens entre l'introduction des nouvelles technologies et l'organisation du temps de travail"* (Consolidated Report by CEGOS), 1985

— **European Foundation for the Improvement of Living and Working Conditions:** *"Participation Review. A Review of Foundation Studies on Participation"* P. Cressey, M. Bolle De Bal, T. Trev, K. Traynor and V. Di Martino), 1988

— **European Foundation for the Improvement of Living and Working Conditions:** *"New Information Technology and Participation in Europe: The Potential for Vocal Dialogue* by Fröhlich; D. Fuchs; H. Krieger), 1989.

Information booklet series:

No. 1 Visual display unit workplaces: emerging trends and problems

No. 2 Safety in hazardous wastes

No. 3 Providing information about urban services

No. 4 Living conditions in urban Europe

No. 5 Commuting in the European Community

No. 6 The reporting of occupational accidents and diseases in the European Community

No. 7 Participation in technological change

No. 8 Adapting shiftwork arrangements. Why and How?

No. 9 New technology in manufacturing industry

No. 10 Office automation

No. 11 Participation in change: New technology and the role of employee involvement

No. 12 Taking action against long-term unemployment in Europe

No. 13 Working for a better environment: The role of the Social Partners

No. 14 Growing up and leaving home

The above, and all Foundation publications, are available upon request from the Offical Sales Offices of the European Communities, the addresses of which are listed at the end of this publication.